Earth's smallest SUPERHEROES

AF126718

PLANKTON

by Ruth Owen

With thanks to:
Professor Abigail McQuatters-Gollop
Professor of Marine Conservation
School of Biological and Marine Science,
University of Plymouth, Plymouth, Devon, UK

Ruby Tuesday Books

Published in 2025 by Ruby Tuesday Books Ltd.

Copyright © 2025 Ruby Tuesday Books Ltd.

All rights reserved. No part of this publication may be reproduced in whole or in part, stored in any retrieval system, or transmitted in any form or by any means, electronic, mechanical, photocopying, recording, or otherwise, without written permission from the publisher.

Editor: Mark J. Sachner
Design & Production: Alix Wood

Picture credits:
Alamy: 4–5 (RGB Ventures), 15T (Blue Planet Archive), 17BL (Panther Media GmbH), 23TL (The Natural History Museum), 23BL (Science Photo Library), 29R (David Fleetham); Creative Commons: 23R, 24B; Nature Picture Library: 13B (Solvin Zankl), 14T (Doug Perrine/Solvin Zankl), 17T (Richard Hermann), 19BL (Franco Banfi), 26T (John Cancalosi); Ruby Tuesday Books: 7TL; Andrei Savitsky: 7BR; Science Photo Library: 7BL (Pascal Goetgheluck), 8C (Karl Gaff), 9B (Choksawatdikorn), 11TL (Gerd Guenther), 11B (Steve Gschmeissner), 13TL (Steve Gschmeissner), 13TR (Steve Gschmeissner), 21BL (Claire Ting), 24T (Wim Van Egmond); Shutterstock: Cover, 4–5 (Napat/Dan Olsen/mitagalihs/Ekky Ilham/lembergarium/shoma81), 4B (BlueOrange Studio), 6T (mitagalihs), 6BL (Andre-Johnson), 6BR (Lebendkulturen.de), 10B (John Hemmings), 14T (Matteo Ciani/Sinhyu Photography), 14B (Rick Beauregard), 15B (VectorMine), 16, 17BR (Mariusz Potocki), 18C (Choksawatdikorn), 18B (Rtimages), 21BR (ixpert), 21C (AOJ17), 21B (sciencepics), 22B (VectorMine), 25T (Narongkan Wanchauy), 26B (Alex Filim), 27T (Romanenkova), 27C (Igor Hotinsky), 27B (Dima Berlin), 28B (Candyfloss Film), 29L (milart); Superstock: 1 (Alex Hyde), 4–5 (Ian Cuming/Ikon–New Paradigm Images), 10T (Ian Cuming/Ikon–New Paradigm Images), 11TR (Alex Hyde), 12T (Hiroya Minakuchi/Minden Pictures), 12B (Richard Hermann/Minden Pictures), 18T (Patricia Robles Gil/Minden Pictures), 19T (Antoine Lorgnier/Biosphoto), 20B (Mark Garlick/SPL); Alix Wood: 3, 22T.

British Library Cataloguing in Publication Data (CIP) is available for this title.

ISBN 978-1-78856-448-9

Printed in Malta by Gutenberg Press Ltd.

www.rubytuesdaybooks.com

Contents

Can Small Save the World? 4
Tiny Drifters 6
The Green Ones 8
Let's Meet Some Phytoplankton 10
Animal Drifters 12
The Zooplankton Babies 14
No Plankton Means Empty Seas 16
Arch Enemy or Sidekick? 18
Plankton Make the Oxygen YOU Breathe ... 20
Earth's Climate Change Heroes 22
When the Good Guys Turn Bad 24
Every Superhero Has Enemies 26
Stand Up for Plankton 28
Glossary 30
Index 32

Can Small Save the World?

This is a book about something small. Something very, very small – plankton!

Aren't plankton just little wiggly things that live in water? I thought this was a book about superheroes.

You're right! Plankton do live in water in oceans, lakes, rivers and ponds.

And plankton are little. In fact, many plankton are so tiny, it's as if they are invisible.

OK. Invisibility is super-cool. But uh-oh! I once swallowed some seawater. Did I swallow plankton?

4

There are trillions and trillions of plankton in the ocean. In some places, just a teaspoon of ocean water will contain thousands of individual plankton. So if you've swallowed seawater, you've probably swallowed plankton!

100,000,000,000,000,

So there are trillions of them. We can't always see them. Hmmm . . .
. . . still finding it hard to get excited.

We get it! Invisible Octopus Woman and Mega-Toothed Shark Man sound more like superheroes. But plankton have powers and abilities that defy our imagination.

Without THESE tiny superheroes, there would be no animals in Earth's oceans. Humans could not survive. And all life on our planet would be in danger!

Wow. OK. Now THAT sounds like a superhero. Please tell me more.

Tiny Drifters

So are plankton animals or plants?

Plankton is a mixture of many different types of living things. They are grouped together as plankton because of how they live.

Bacteria, viruses and fungi that live in water

Microscopic plant-like algae

The eggs and larvae of bigger animals such as fish and octopuses

Microscopic animals

Octopus larva

A water flea with eggs

Most plankton can't or don't swim. They float in water and drift from place to place in ocean currents.

"Plankton" comes from the Greek word for "drifting."

How do scientists capture and study plankton?

Just as you would use a net to dip for little animals in a pond or rock pool, scientists use nets to capture plankton. The nets have a tiny mesh to trap the miniature living things.

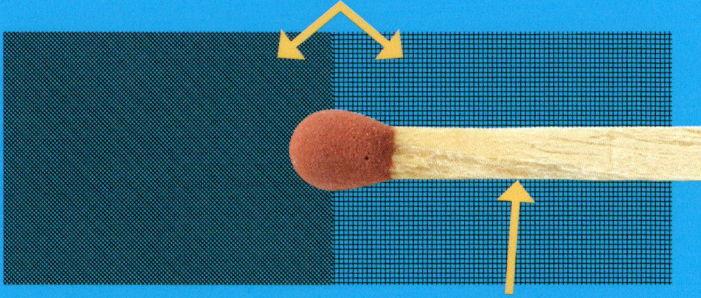

Two sizes of plankton mesh

Matchstick

Scientists look at samples of water that contain plankton under powerful microscopes.

Microscope

Plant-like algae

Plankton don't ONLY drift

Copepods are tiny animals that swim like superheroes. When chasing prey or escaping from predators, they can move through water at 91 metres per hour.

Wait a minute. That doesn't sound very fast.

Well. You have to remember that a copepod may be only 1 millimetre long.

So let's do the maths. Take a copepod's tiny size . . . measure a human . . . multiply that number . . . miles per hour . . .

÷
60
5280 90,000 405,000
×

OK. If you were as fast as a copepod, you could swim at 75 miles (120 km) per hour!

The Green Ones

Scientists divide plankton into two main groups – phytoplankton and zooplankton.

Phytoplankton are microscopic living things. This group includes tiny, plant-like algae. It also includes many types of bacteria.

If you placed five of these tiny phytoplankton in a row, they would measure just 1 millimetre!

This is a type of freshwater phytoplankton, or algae. The green areas are called chloroplasts.

The chloroplasts contain chlorophyll, which makes a phytoplankton look green.

Measuring Plankton

Scientists measure plankton in centimetres, millimetres and microns.

There are 10 millimetres in one centimetre (cm).

1 millimetre

There are 1000 microns in 1 millimetre!

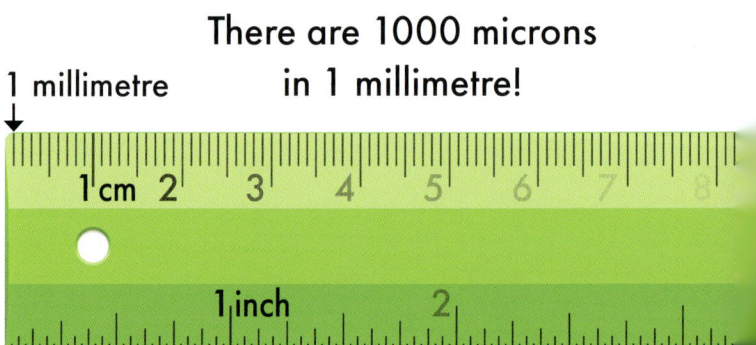

Just like plants, phytoplankton can make their own food for energy and growth. They do this by a process called photosynthesis.

Phytoplankton Photosynthesis

Most phytoplankton float in the top 100 metres of water, where there is most sunlight.

Carbon dioxide gas from the air enters the water.

An algae-type phytoplankton absorbs carbon dioxide gas.

It also absorbs water.

Phytoplankton

Inside its chloroplasts, the phytoplankton uses the Sun's light to turn the water and carbon dioxide into sugary food.

Bacteria phytoplankton don't have chloroplasts, but they are still able to make their food by photosynthesis.

Groups of phytoplankton called cyanobacteria

Scientists estimate there could be one million different kinds of phytoplankton and zooplankton.

Let's Meet Some Phytoplankton

So phytoplankton are really tiny, they can look green and they photosynthesise, right?

Right! But take a look with a microscope, and you'll see that phytoplankton can also grow armour, grow extra parts and even use teamwork to survive.

Coccolithophores

Coccolithophores are ocean phytoplankton that grow rocky armour.

Scientists think the armour may be a defence against other plankton that eat them.

Coccolithophore

Size: 10 microns

The pieces of armour are called coccoliths.

When coccolithophores die, their armour crumbles and becomes sand.

Many rocks, such as the White Cliffs of Dover in England, formed from the armour of prehistoric coccolithophores.

Diatoms

Diatoms form a protective glass-like shell.

Different-shaped diatoms

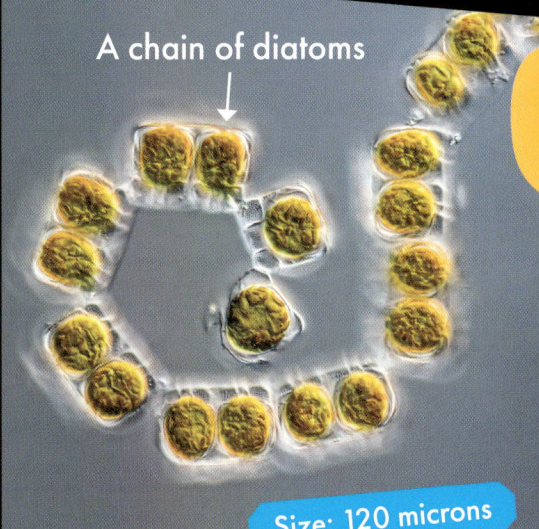

A chain of diatoms

Some types of diatoms use sticky mucus to join up and make chains.

This teamwork helps them float and stay close to the water's surface, where there is plenty of sunlight.

Size: 120 microns

Dinoflagellates

Dinoflagellates live in oceans and in freshwater. They have two parts called flagella that help them move through water.

One type of dinoflagellate grows fingers filled with chloroplasts each day. The fingers absorb sunlight for photosynthesis. Then, at sunset, the dinoflagellate retracts the fingers.

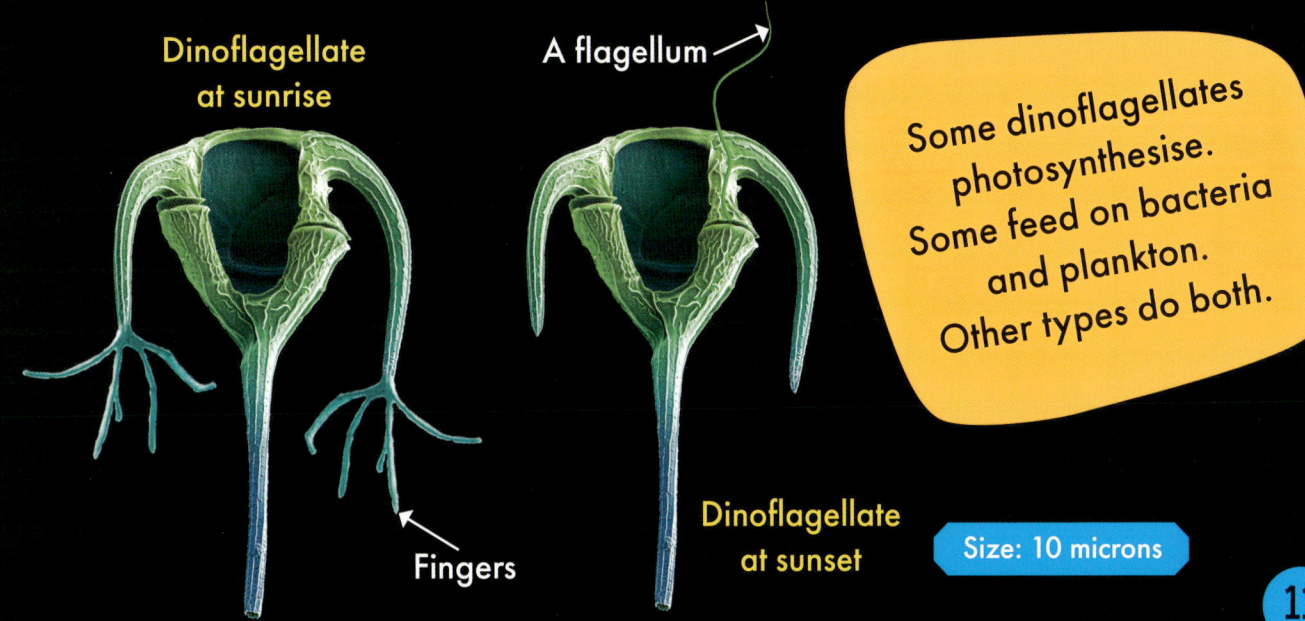

Dinoflagellate at sunrise

A flagellum

Fingers

Dinoflagellate at sunset

Some dinoflagellates photosynthesise. Some feed on bacteria and plankton. Other types do both.

Size: 10 microns

Animal Drifters

Zooplankton are tiny animals that drift in oceans and freshwater.

Most zooplankton spend their days in deep, dark water avoiding birds, fish and other predators. At night, they move up to the water's surface to feed.

The word "zoo" comes from the Greek word zoios, which means "animal".

Krill

These zooplankton form swarms that are so huge they can be seen from space!

Krill

Size: 6 cm

Krill are relatives of crabs, lobsters and shrimps.

A swarm of krill →

← Scientist

Krill feed on phytoplankton. However, if food is scarce, they can survive for 200 days without eating.

Foraminifera

A foraminifera is a plankton that forms a skeleton called a test. The test is covered with tiny holes.

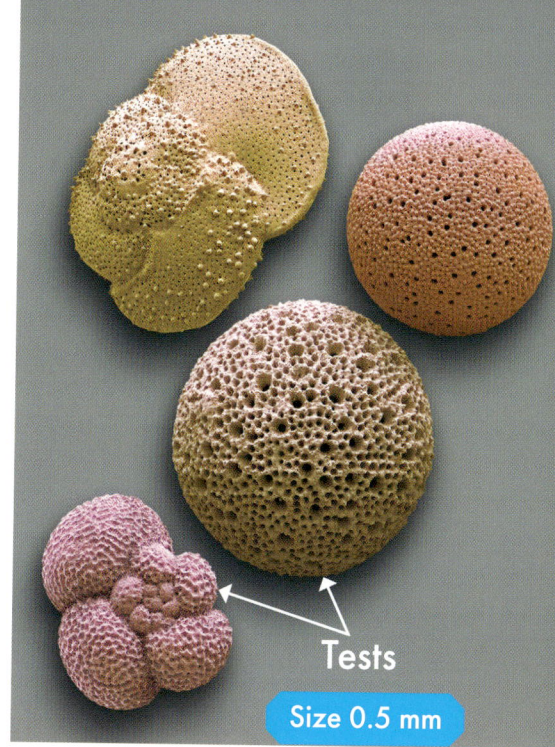

Tests

Size 0.5 mm

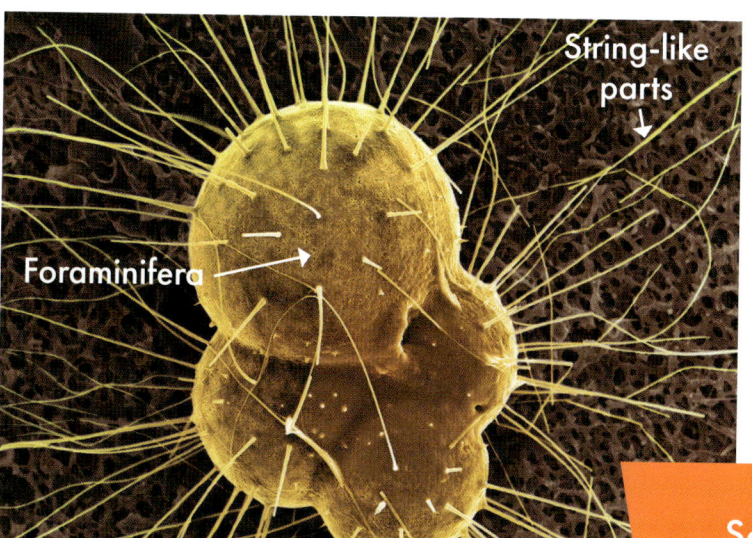

Thin, string-like parts of the plankton's body reach through the holes to capture food.

Some types of foraminifera feed on phytoplankton. Some feed on zooplankton. Others catch and eat waste such as floating bits of dead animals or seaweed.

Phronima

This ferocious zooplankton hunts for small, jelly-like animals such as salps.

The phronima eats the salp's insides. Then it lives in its victim's empty jelly-like barrel, or body.

A female phronima's larvae hatch from their eggs and live with their mother in her floating home.

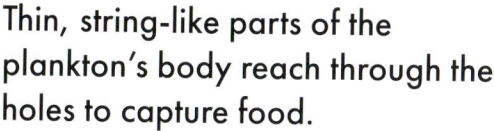

Size: 2.5 cm

The Zooplankton Babies

Lobster larva

Some animals only live as zooplankton when they are babies.

Then these tiny, drifting larvae grow up to become bigger animals such as fish, crabs and octopuses. As long as they are not eaten first!

Crab larva

To tiny, soft-bodied planktonic animals, water feels very thick. They drift because swimming would be as difficult as you trying to swim through treacle.

Fish larva

Starfish larva

Baby Barnacles

When baby barnacles first hatch from their eggs, they live as microscopic zooplankton.

Once they are ready to begin their adult lives, they attach to rocks and grow shells.

Barnacle larvae

Adult barnacles

A Million-Year-Old Jellyfish?

There's one plankton animal that scientists believe may have the superhero skill of living forever – the immortal jellyfish!

If this tiny jellyfish is injured or starving from lack of food, it simply goes back in time.

It attaches itself to the seabed and goes back to the polyp stage of its life cycle. Then it begins its adult life again.

An immortal jellyfish measures just 3 mm across.

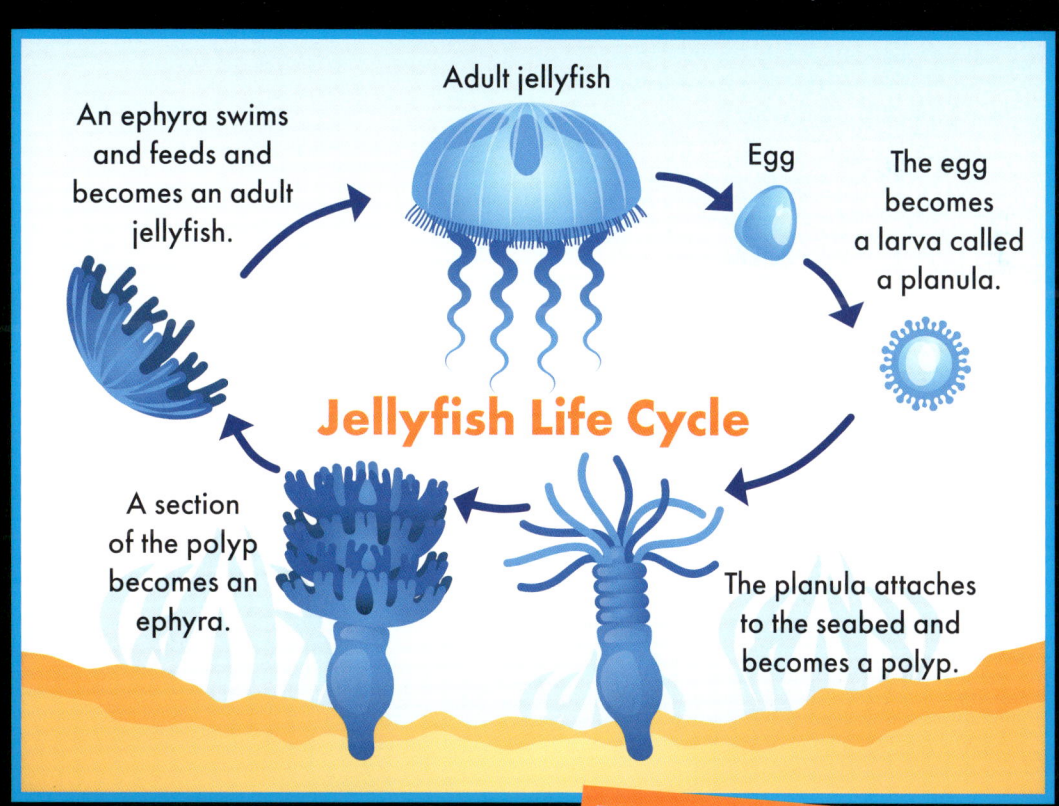

Jellyfish Life Cycle

- Adult jellyfish
- Egg
- The egg becomes a larva called a planula.
- The planula attaches to the seabed and becomes a polyp.
- A section of the polyp becomes an ephyra.
- An ephyra swims and feeds and becomes an adult jellyfish.

Could there be immortal jellyfish in the ocean that have been alive for millions of years? No one can say for sure!

WOW! Plankton really are super-cool. So how can plankton help save the world?

No Plankton Means Empty Seas

Every animal that lives in the ocean needs plankton to survive.

Not a great white shark? It's much too cool to eat tiny plankton.

Wrong! Without plankton, even sharks would become extinct.

Sunshine → Phytoplankton → Zooplankton → Fish → Seal → Shark

Every ocean food chain (above) or food web (below) begins with sunshine and phytoplankton.

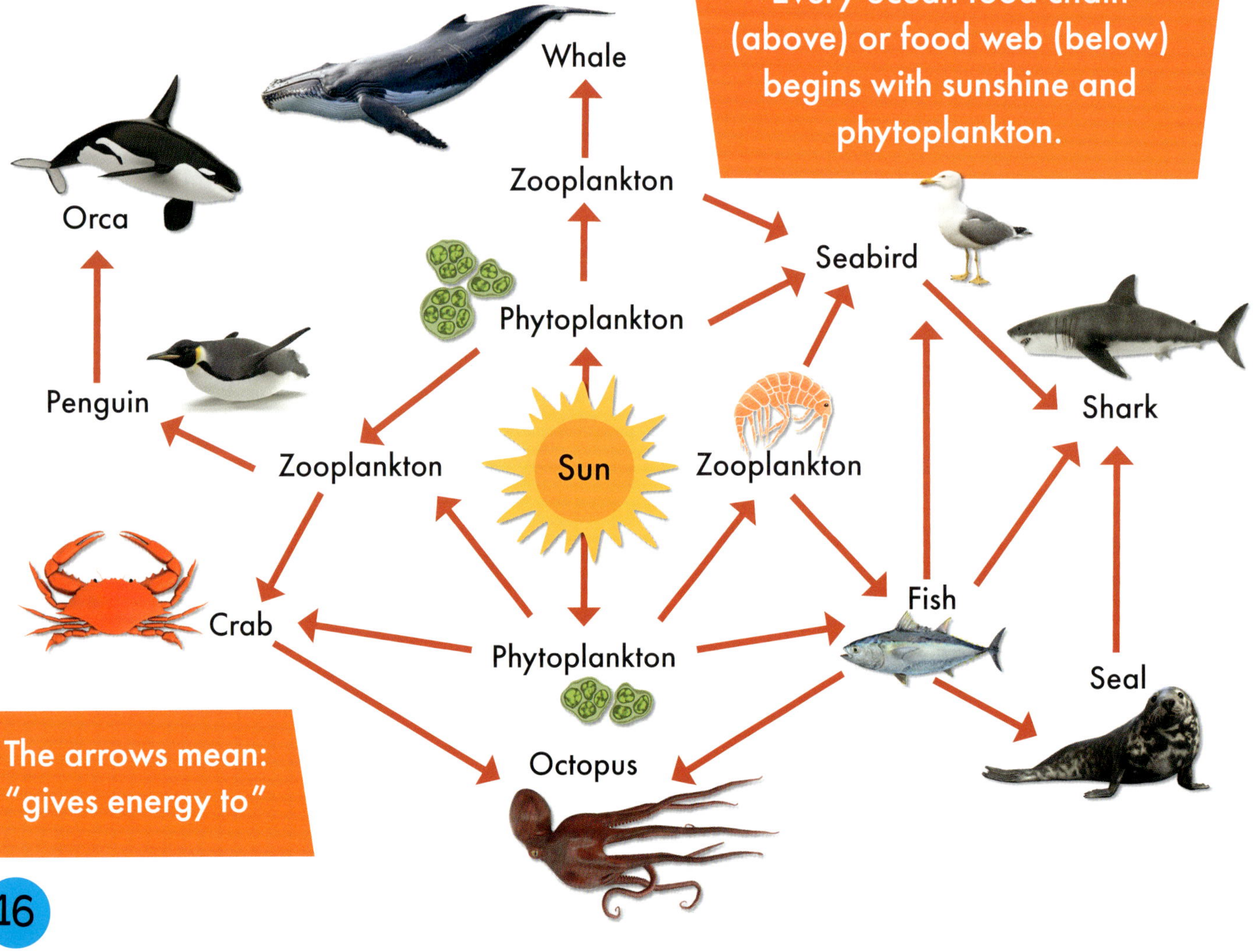

The arrows mean: "gives energy to"

16

For many ocean animals, including whales, seals, penguins and fish, krill are an essential food.

Krill swarm

The word "krill" means "whale food" in Norwegian.

Blue Whales

A blue whale opens its mouth wide, swallowing water and vast quantities of krill.

Blue whale

One scientific study discovered that a blue whale can eat about 14.5 tonnes of krill in a day. That's the weight of a bus filled with passengers!

Crabeater Seals

You might think that crabeater seals eat crabs. Wrong! A crabeater seal scoops up mouthfuls of krill. Its teeth trap the krill and the seawater strains out.

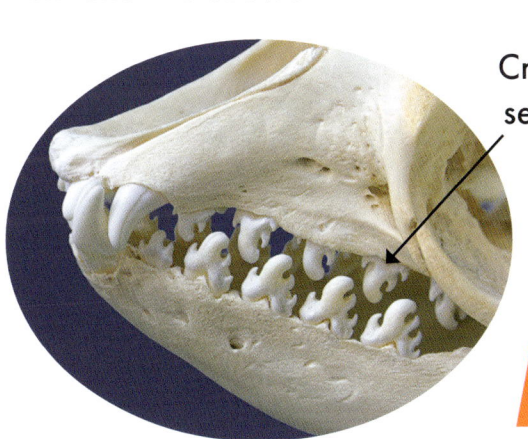

Crabeater seal teeth

Crabeater seal

A crabeater seal may eat up to 11,000 krill in a day!

Arch Enemy or Sidekick?

Millions of plankton are eaten by whales every day. So you might think whales are the arch enemies of our tiny superheroes.

However, like the best superhero stories, here comes a twist....

A scientist studying krill

Whales are in fact trusty sidekicks to ocean plankton. Why?

Phytoplankton can use sunshine to make the food they need for energy. But they also need nutrients such as iron, nitrogen and calcium.

So how do phytoplankton get nutrients?

Phytoplankton

You might eat a sandwich or a big bowl of pasta for energy. But you also need nutrients from foods such as broccoli and peas.

I don't like broccoli and peas.

Sorry! We don't make the rules.

Phytoplankton can get nutrients from **WHALE POO**.

A whale, such as a humpback whale, eats massive quantities of krill and other zooplankton.

When it needs to go, the whale releases huge amounts of nutrient-rich poo into the water.

A humpback whale scooping up water and plankton

Phytoplankton absorb the nutrients from the water and use them to grow. Then zooplankton eat the phytoplankton.

Floating whale poo

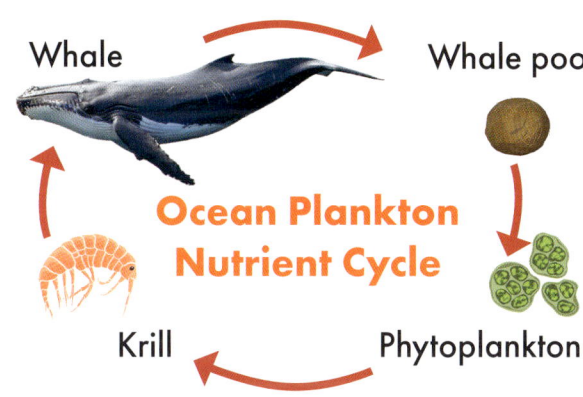

Whale → Whale poo → Phytoplankton → Krill → Whale

Ocean Plankton Nutrient Cycle

Whales need plankton as food. And plankton need whales to get nutrients. THAT's some superhero teamwork!

Plankton Make the Oxygen YOU Breathe

What? I thought rainforests make oxygen.

That's correct. Rainforests do make oxygen, but so do phytoplankton. In fact, phytoplankton make 50% of the oxygen in Earth's atmosphere!

Just like rainforest trees and other land plants, phytoplankton make oxygen as they photosynthesise. Then they release the oxygen into the air.

Trees and other land plants

Oxygen

Oxygen

Phytoplankton

Making Life Possible

About 3.5 billion years ago, the first prehistoric ocean formed on Earth. It was home to microscopic plankton.

The plankton began to photosynthesise and make oxygen – billions of years before land plants existed.

These ancient plankton helped make life on Earth possible!

This place is rubbish!

Let's make it green and habitable.

Prochlorococcus

Prochlorococcus is a cyanobacteria that lives in oceans. It produces up to 20% of Earth's oxygen!

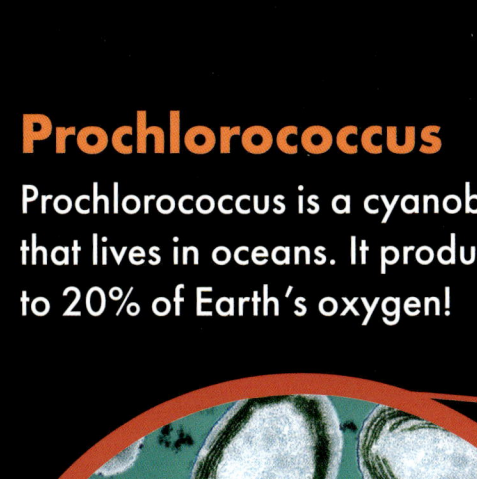

Up to 0.7 microns across

Prochlorococcus seen through a powerful microscope

Scientists only discovered this microscopic phytoplankton in the late 1980s.

A single drop of water can contain 20,000 of this cyanobacteria.

Prochlorococcus can survive in warm or cold waters. It can even photosynthesise and make oxygen in deep water with very little light.

So take two deep breaths and thank Earth's phytoplankton for one of them.

You're welcome!

Earth's Climate Change Heroes

Superheroes fight bad guys, right? So who do plankton fight?

Earth's plankton are superheroes who help us fight climate change!

What do we want? Less carbon dioxide in the air. When do we want it? Now! Walk don't drive! Switch off that light!

Earth is getting too warm because of harmful gases in its atmosphere. One of these gases is carbon dioxide.

Every year, phytoplankton remove HUGE quantities of carbon dioxide from the air when they photosynthesise.

Climate Change Fast Facts

Climate change is happening because of human activities.

Cars, trucks and planes release gases when they burn fuel.

When a power station burns coal to make electricity, harmful gases are released into Earth's atmosphere.

The gases, which include carbon dioxide, trap too much of the Sun's heat, making Earth get warmer.

When phytoplankton take in carbon dioxide from the air, it stays inside them.

Coccolithophore

Coccolithophores use carbon to make their armour.

Once a phytoplankton dies, it sinks to the seabed or bottom of a lake or pond. The carbon goes, too, and is trapped under the water and out of Earth's atmosphere!

The armour of dead coccolithophores on the seabed

When zooplankton eat phytoplankton, the carbon enters their bodies and is removed from the atmosphere.

Zooplankton Are Climate Change Heroes

Many zooplankton animals, such as krill and water fleas, have shells, or exoskeletons.

When they eat phytoplankton, carbon from their food is stored in their shells.

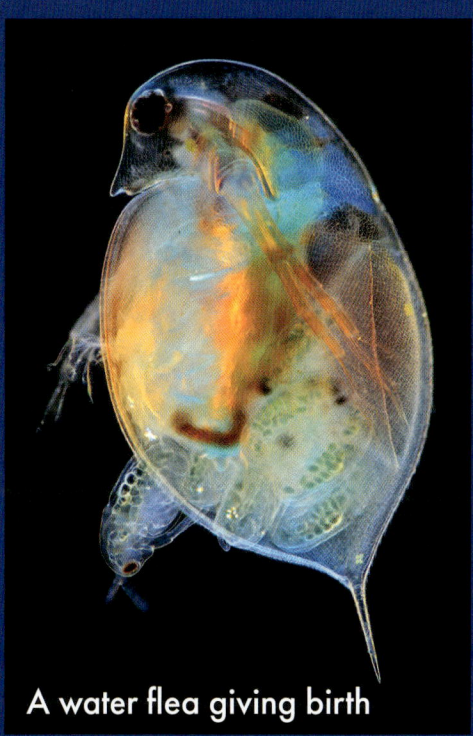

A water flea giving birth

These animals regularly shed their shells and grow new ones.

The old shells fall to the seabed or to the bottom of ponds and lakes. Billions of these shells are underwater, storing carbon.

When the Good Guys Turn Bad...

Sometimes, the numbers of phytoplankton in one part of the ocean or in a lake or pond increase too much. This is called a bloom.

Most types of phytoplankton reproduce by dividing in half – one becomes two!

 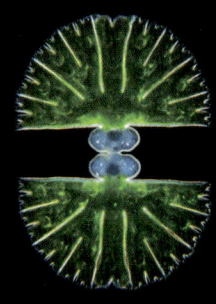 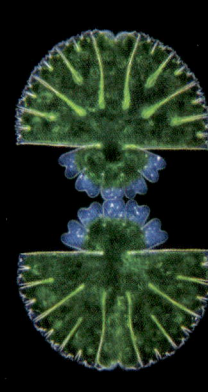 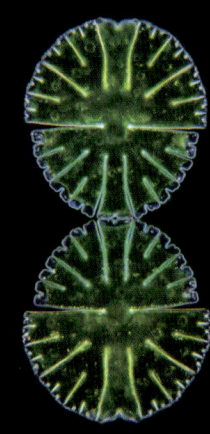

Large numbers of phytoplankton can be a good thing. It can mean more food for other living things in that water.

Sometimes, however, a phytoplankton bloom can cause terrible harm.

The pale greenish-blue areas are a phytoplankton bloom.

Lake Erie seen from space

During a plankton bloom, the huge amount of plankton form a terrible, smelly, green, red or brown slime on the water.

Plankton bloom

A poisoned fish

A plankton bloom blocks out sunlight. Other living things can't survive in the darkness.

Phytoplankton make oxygen, but they also use oxygen to live. A huge quantity of phytoplankton can remove all the oxygen from the water. Fish, insects and other water animals suffocate.

Water becomes poisoned, harming swimmers or contaminating drinking water.

TOXIC

A plankton bloom produces toxins that contaminate the water. Harmful gases are also released into the air.

Fish and shellfish can take in the toxins, be poisoned, and become dangerous for people to eat.

But I thought plankton were the good guys. Why do they go bad?

Every Superhero Has Enemies

Sadly, when plankton go bad, it is often because of human actions!

Phytoplankton need nutrients to grow and reproduce. However, if too many nutrients get into the water, plankton reproduce too much.

A snapping turtle fights to survive in a plankton bloom.

So how do unwanted nutrients get into water?

Nutrient-rich fertilisers that are used to feed garden plants or farm crops can mix with rain. So can other chemicals or oil. The rain then runs down storm drains or into rivers. Eventually, the nutrient-rich water flows into the ocean.

Storm drain

Oily chemicals

Human sewage from waste treatment plants is filled with nutrients. If this escapes into rivers, lakes or oceans, the nutrients go, too.

Oil that escapes from tanker ships or oil drilling rigs can help feed phytoplankton.

Oil drilling rig

High levels of carbon dioxide in the air from power stations, factories and vehicles help phytoplankton photosynthesise and increase in number.

Just like everything to do with nature and our planet's health, balance is essential.

With the right amount of nutrients, plankton are the good guys. But too many nutrients can turn them bad!

Stand Up for Plankton

Plankton are SO awesome! What can we do to help them?

Scientists and conservationists around the world are spreading the word about the importance of plankton. You can, too!

Never allow litter to get into rivers, lakes and the ocean. When we take care of water, we take care of plankton.

Ask your family not to use chemical fertilisers in your yard.

Do everything YOU can to help fight climate change.

Governments must stop companies from polluting oceans, lakes and rivers with fertilisers, other chemicals, oil or sewage.

Plankton are Amazing!

Some types of phytoplankton contain substances that can create a blue light called bioluminescence.

Plankton bioluminescence

Plankton are Amazing!

Radiolarian are zooplankton that team up to live in colonies. They surround themselves with a jelly-like substance.

Zooplankton shrimp

Radiolarian

Jelly

The time has come to care about plankton in the same way that we care about Earth's biggest treasures, such as whales and giant rainforest trees.

In the future, phytoplankton could be grown as food for humans and farm animals. This could mean fewer forests are cut down to make space to plant crops.

Spirulina is a freshwater cyanobacteria. It is made into a powder or pills that people eat to get extra nutrients.

Scientists are studying plankton that could help fight infections, viruses and even cancer!

We can all join these tiny good guys in their battle to protect planet Earth.

Join Team Plankton today.

Because when we stand up for plankton, we protect every other living thing on Earth!

Glossary

algae

Plant-like living things that make their own food by photosynthesis. Most types of algae live in water. Seaweeds are algae, and so are many types of phytoplankton.

bacteria

Microscopic living things that are also known as microbes. Some bacteria (germs) can cause disease. Many types of bacteria live in oceans, lakes, rivers and ponds as plankton.

carbon dioxide

A colourless gas in the air that is a natural part of Earth's atmosphere. Plants, seaweeds and phytoplankton take in carbon dioxide when they photosynthesise. Humans and animals release carbon dioxide when they breathe out. When fuels, such as oil and coal, are burned, carbon dioxide is released into the air.

chloroplasts

Tiny parts inside plants, seaweeds and many types of phytoplankton where photosynthesis takes place. The chlorophyll in chloroplasts make phytoplankton and land plants look green.

climate change

A change in Earth's temperature and weather that happens over a long period of time. Climate change can happen naturally, but at the moment, Earth's temperature is getting warmer primarily because of human actions.

cyanobacteria

Bacteria that are able to photosynthesise. Many live in water as plankton.

drift

To be carried, or moved, slowly by water or air currents.

immortal

Able to live forever.

larvae

The young of many water animals, such as fish and crabs, and land animals, such as insects. A larva hatches from an egg.

micron

A tiny unit of measurement. There are 1000 microns in a millimetre. There are 10,000 microns in a centimetre.

microscopic
Only visible through a microscope.

nutrients
Substances that are needed by living things for health and growth. For example, nitrogen, calcium and vitamin C are all types of nutrients.

oxygen
A colourless gas in the air that is a natural part of Earth's atmosphere. Humans and animals need oxygen to breathe. Plants, seaweeds and phytoplankton make and release oxygen during photosynthesis.

photosynthesis
The process during which sunlight is used to turn water and carbon dioxide gas into a sugary food for energy and growth. Plants, seaweeds and phytoplankton are able to photosynthesise.

phytoplankton
A group of plankton that includes microscopic plant-like algae and many types of bacteria. Phytoplankton make their own food by photosynthesis.

plankton
Living things such as bacteria, microscopic animals and microscopic plant-like algae that live in water. They are grouped together as plankton because they mostly float or drift from place to place in water currents.

zooplankton
A group of plankton that includes microscopic animals and the tiny larvae, or young, of larger water animals. Some types feed on phytoplankton, some feed on other zooplankton, some feed on both phytoplankton and each other and some types feed on waste.

Index

A
algae 6–7, 8–9
animals 6–7, 8–9, 12–13, 14–15, 16–17, 18–19, 23, 25, 26, 29

B
bacteria 6, 8–9, 11, 21, 29
barnacles 14
bioluminescence 28
blooms 24–25

C
carbon dioxide 9, 22–23, 27
climate change 22–23
coccolithophores 10, 23
copepods 7
cyanobacteria 9, 21, 29

D
diatoms 11
dinoflagellates 11

F
foraminifera 13

I
immortal jellyfish 15

K
krill 12, 17, 18–19, 23

L
larvae 6, 13, 14–15

M
measuring plankton 8
microscopes 7, 10, 21

O
oxygen 20–21, 25

P
photosynthesis 9, 10–11, 16, 18, 20–21, 22, 27
phronima 13
phytoplankton 6–7, 8–9, 10–11, 12–13, 16, 18–19, 20–21, 22–23, 24–25, 26–27, 28–29
prochlorococcus 21

R
radiolarian 29

S
seals 16–17
spirulina 29
studying plankton 7, 18

V
viruses 6, 29

W
water fleas 6, 23
whales 16–17, 18–19, 29

Z
zooplankton 6–7, 8–9, 12–13, 14–15, 16–17, 19, 23, 29